The Reason: Other Side

By

Ian Beardsley

ISBN: 978-1-312-33466-3

Clearing a few things up about The Reason, The Reason: Moving On, and Sidenotes For The Reason

For the Sagittarius vector, we took the parameter, t, to be zero. For the Scutum vector, we used the same three equations, Neptune, Uranus, Earth, to make an equation of the plane, but we did not set t = 0, but rather eliminated t. However we reflected the declination about the equator by saying it was negative 10 degrees instead of 10 degrees. If you leave it as ten degrees, the vector points to an open cluster in Aquila, NGC 6738. When we derived the Pegasus vector, we switched the x coordinate with the y coordinate, essentially reorienting the coordinate system. The conventional use of

a coordinated system comes from random definition anyway, so we don't know which an extraterrestrial wants us to use.

We have

$$\frac{5x}{36} + \frac{20}{36} = t$$

$$\frac{10y}{33} - \frac{30}{33} = t$$

$$\frac{z}{9} + \frac{5}{9} = t$$

Set $t=0$:

$$\frac{5x}{36} + \frac{20}{36} = \frac{10y}{33} - \frac{30}{33}$$

$$\frac{5x}{36} - \frac{10y}{33} + \frac{20}{36} + \frac{30}{33} = \frac{z}{9} + \frac{5}{9}$$

$$\frac{5x}{36} - \frac{10y}{33} + \frac{145}{99} = \frac{z}{9} + \frac{5}{9}$$

$$\frac{5x}{36} - \frac{10y}{33} + \frac{145}{99} - \frac{z}{9} - \frac{5}{9} = 0$$

$$\frac{5x}{36} - \frac{10y}{33} - \frac{z}{9} + \frac{10}{11} = 0$$

$$\vec{F} = \langle \frac{5}{36}, -\frac{10}{33}, -\frac{z}{9} \rangle$$

This points to sagittarius
near the wow! signal as
calculated in The Reason by Ian Beardsley

4

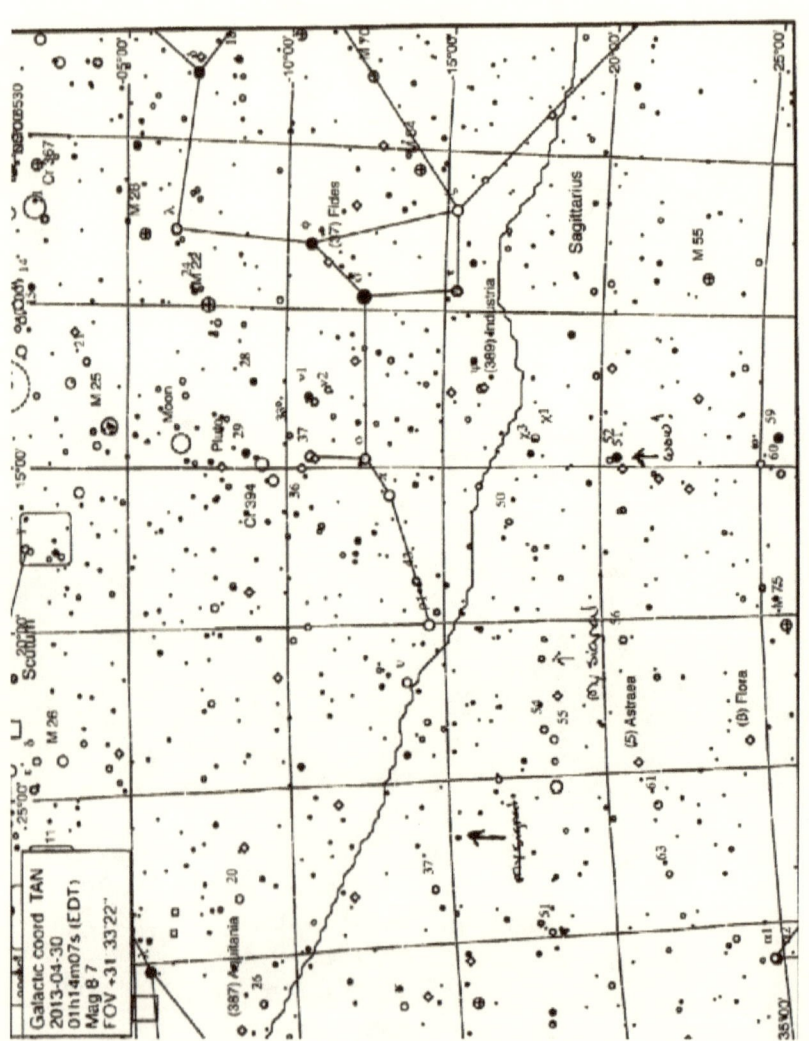

we wrote in sidenotes for The Reason

$$\nabla f = \langle \tfrac{5}{36}, -\tfrac{20}{33}, -\tfrac{1}{9} \rangle$$

~~should~~ would be otherwise written.

$$\nabla f = \langle \tfrac{5}{36}, -\tfrac{20}{33}, \tfrac{1}{9} \rangle$$

by eliminating t in:

~~t = 5x + 20~~

$$t = \tfrac{5x}{36} + 20 \qquad t = \tfrac{10y}{33} - 30 \qquad t = \tfrac{7}{9} + 5$$

This Gives:

RA: 19 hours

Dec: +10°

This is the ~~st~~ around the ~~stars~~

~~constellation~~ open cluster

NGC 6738 in Aquila

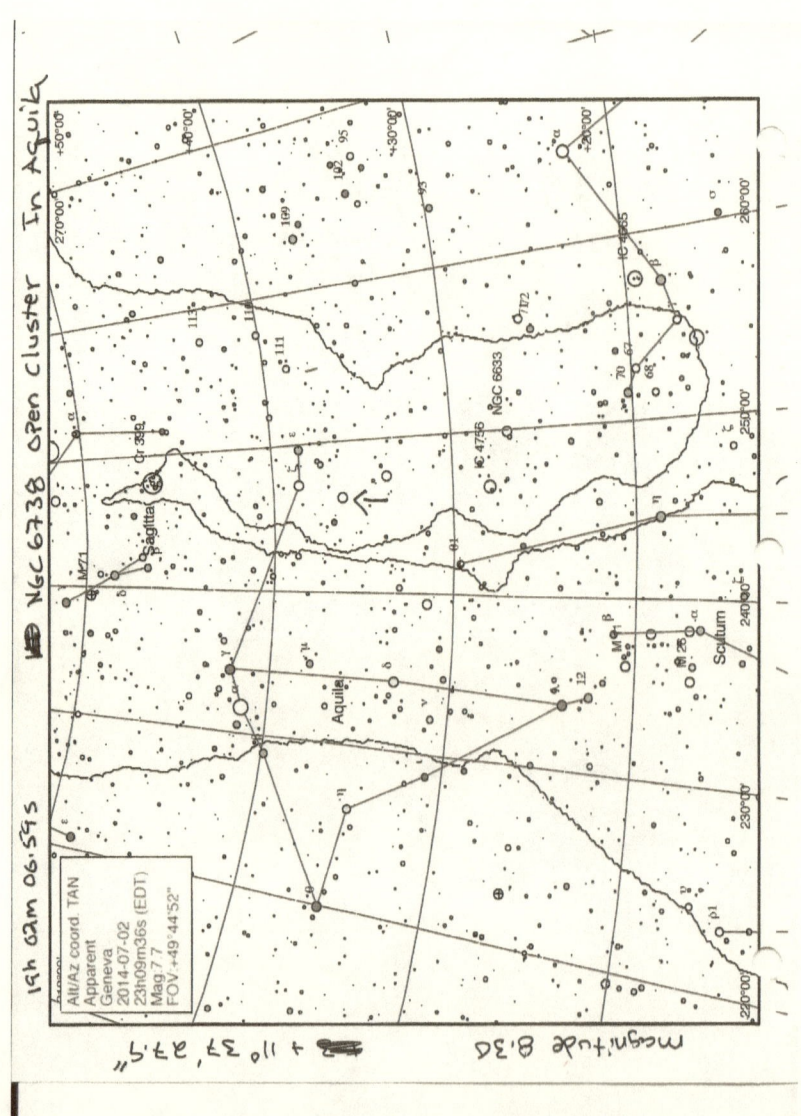

6

NGC 6738 Open Cluster In Aquila

19h 02m 06.59s

Magnitude 8.39   + 11° 37' 22.6"

From The Reason by Ian Beardsley

$(\frac{9}{5}, \frac{5}{3}, 2) = A$

This is around the Star HD 180938 near α Vul

$(1, 4, 7) = B$

$A \times B = \langle \frac{11}{3}, -\frac{53}{5}, \frac{83}{15} \rangle$

$a = \frac{11}{3}$ $b = -\frac{53}{5}$ $d = \frac{83}{15}$

$c^2 = (\frac{11}{3})^2 + (\frac{53}{5})^2$ $c = 11.2$

$\tan \alpha = \frac{b}{a} =$ ~~162456 06-2026.0605 RA~~

~~~~ $\tan a = \frac{b}{a} = -\frac{53}{5} \div \frac{11}{3}$

$= -\frac{159}{55} = 2.89$

$\alpha = -70.91°$

In The Reason we already said $B = 26° 17' 20.4''$

$\alpha = RA$ $B = Dec$, $\frac{-70.91}{15 \, deg/hr} = -4.7273$ hours

$(0.7273 \, hr)(60 \, min/hr) = 43.638$ min

$(0.638 \, min)(60 \, sec/min) = 38.28''$

$24^h 00^m 00^s - 4.7273^h = 19.2727^h$

$(0.2727)(60) = 16.362^m$ $(0.362^m)(60) = 21.72''$

our star is at: $RA = 19^h 16^m 21.72''$

$Dec = 26° 17' 20.4''$

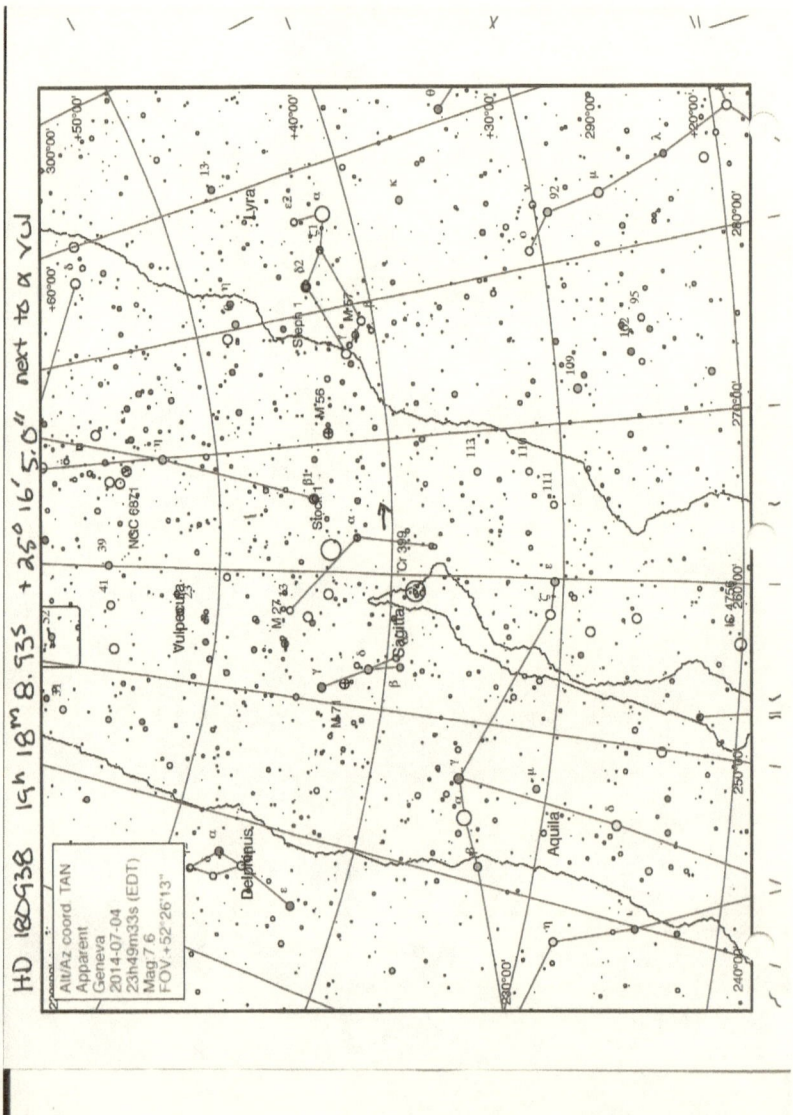

HD 180938 19ʰ 18ᵐ 8.73ˢ + 26° 16' 5.0" next to α Vul

That our $9/5$, $5/3$, 2,... have normal vector pointing to the
Constellation Vulpecula is significant because, as Wikipedia
writes:

Vulpecula is also home to HD 189733 b, one of the closest extrasolar
planet currently being studied by the Spitzer Space Telescope. On 12
July 2007 the *Financial Times* (London) reported that the chemical
signature of water vapour was detected in the atmosphere of this
planet. Although HD 189733b with atmospheric temperatures rising
above 1,000 °C is far from being habitable, this finding increases the
likelihood that water, an essential component of life, would be found
on a more Earth-like planet in the future.

Honing In On The AE-35

My search for what the essence of the AE-35 antenna is, that concept wherein a Gypsy shaman made me telepathic with him, has finally been realized: I actually knew two Manuel's, not just two by the same name, but identical in spirit, and diametric opposites in personality; one a shaman at one end of the Sacromonte, the other at the other end of the Sacromonte. Never did I know them to be at the same place at the same time, or even to be aware one of the other. It in fact could be it is one person who can be at two different places at once. What we are dealing with here, in terms of physics, is quantum teleportation, or quantum entanglement: It is stated a particle can instantaneously exist in another place, and whether the teleported particle actually jumped from one place to another instantaneously or one particle vanished with a separate, or identical one, being reborn in another place makes no difference. Also it is entanglement: Two particles collide setting up initial conditions, as they travel apart each behaves according to the state of the other suggesting instantaneous communication between them over distance even though this violates the cosmic speed limit of the speed of light. Luckily, there is some indication that Manuel will be giving a lecture, In Spain, about the AE-35 Antenna. It is my hope that it will be on youtube.

Building The Antenna

I have said around 1990 that Manuel had me help him install an antenna for television reception at his nephew's house in the Sacromonte (Gypsy Caves). In the act of doing this I became telepathic with him. We call it the AE-35 antenna from 2001: A Space Odyssey. However the extraterrestrials undermined the spell in a plan that was initiated even before I knew Manuel. In 1977 they sent a transmission from the direction of the constellation Sagittarius that was received by the Search For Extraterrestial Intelligence (SETI). It was The Gypsy Shamanism of Manuel that lead me to the discovery that the signal should repeat itself on August 15, 2015. This gave me the idea of building a radio telescope, essentially an antenna not unlike the one I helped Manuel install, so I could receive the predicted message. It is July 05, 2014 now. This gives me a little over a year to build the Antenna, mount it and put in a tracking drive, and to interface it with my computer, as well as find software that will process the signal. The act of building this antenna should not subvert Manuel's spell, but alter its agenda so that Manuel, I, and the extraterrestrials are all working together telepathically with a new proposed agenda, my guess rooted in the concept of welcoming humanity into the stellar community. I have a toolbox with electrical components and electronics tools that I have put together over the years to further my education in electronics, if not to find work in electrical engineering.

The Toolbox

Chapter 1

AE-35

I wrote a short story last night, called Gypsy Shamanism and the Universe about the AE-35 unit, which is the unit in the movie and book 2001: A Space Odyssey that HAL reports will fail and discontinue communication to Earth. I decided to read the passage dealing with the event in 2001 and HAL, the ship computer, reports it will fail in within 72 hours. Strange, because Venus is the source of 7.2 in my Neptune equation and represents failure, where Mars represents success.

Ian Beardsley
August 5, 2012

It must have been 1989 or 1990 when I took a leave of absence from The University Of Oregon, studying Spanish, Physics, and working at the state observatory in Oregon -- Pine Mountain Observatory—to pursue flamenco in Spain.

The Moors, who carved caves into the hills for residence when they were building the Alhambra Castle on the hill facing them, abandoned them before the Gypsies, or Roma, had arrived there in Granada Spain. The Gypsies were resourceful enough to stucco and tile the abandoned caves, and take them up for homes.

Living in one such cave owned by a gypsy shaman, was really not a down and out situation, as these homes had plumbing and gas cooking units that ran off bottles of propane. It was really comparable to living in a Native American adobe home in New Mexico.

Of course living in such a place came with responsibilities, and that included watering its gardens. The Shaman told me: "Water the flowers, and, when you are done, roll up the hose and put it in the cave, or it will get stolen". I had studied Castilian Spanish in college and as such a hose is "una manguera", but the Shaman called it "una goma" and goma translates as rubber. Roll up the hose and put it away when you are done with it: good advice!

So, I water the flowers, rollup the hose and put it away. The Shaman comes to the cave the next day and tells me I didn't roll up the hose and put it away, so it got stolen, and that I had to buy him a new one.

He comes by the cave a few days later, wakes me up asks me to accompany him out of The Sacromonte, to some place between there and the old Arabic city, Albaicin, to buy him a new hose.

It wasn't a far walk at all, the equivalent of a few city blocks from the caves. We get to the store, which was a counter facing the street, not one that you could enter. He says to the man behind the counter, give me 5 meters of hose. The man behind the counter pulled off five meters of hose from the spindle, and cut the hose to that length. He stated a value in pesetas, maybe 800, or so, (about eight dollars at the time) and the Shaman told me to give that amount to the man behind the counter, who was Spanish. I paid the man, and we left.

I carried the hose, and the Shaman walked along side me until we arrived at his cave where I was staying. We entered the cave stopped at the walk way between living room and kitchen, and he said: "follow me". We went through a tunnel that had about three chambers in the cave, and entered one on our right as we were heading in, and we stopped and before me was a collection of what I estimated to be fifteen rubber hoses sitting on ground. The Shaman told me to set the one I had just bought him on the floor with the others. I did, and we left the chamber, and he left the cave, and I retreated to a couch in the cave living room.

Chapter 2

Gypsies have a way of knowing things about a person, whether or not one discloses it to them in words, and The Shaman was aware that I not only worked in Astronomy, but that my work in astronomy involved knowing and doing electronics.

So, maybe a week or two after I had bought him a hose, he came to his cave where I was staying, and asked me if I would be able to install an antenna for television at an apartment where his nephew lived.

So this time I was not carrying a hose through The Sacromonte, but an antenna.

There were several of us on the patio, on a hill adjacent to the apartment of The Shaman's Nephew, installing an antenna for television reception.

Chapter 3

I am now in Southern California, at the house of my mother, it is late at night, she is a asleep, and I am about 24 years old and I decide to look out the window, east, across The Atlantic, to Spain. Immediately I see the Shaman, in his living room, where I had eaten a bowl of the Gypsy soup called Puchero, and I hear the word Antenna. I now realize when I installed the antenna, I had become one, and was receiving messages from the Shaman.

The Shaman's Children were flamenco guitarists, and I learned from them, to play the guitar. I am now playing flamenco, with instructions from the shaman to put the gypsy space program into my music. I realize I am not just any antenna, but the AE35 that malfunctioned aboard The Discovery just before it arrived at the planet Jupiter in Arthur C. Clarke's and Stanley Kubrick's "2001: A Space Odyssey". The Shaman tells me, telepathically, that this time the mission won't fail.

Chapter 4

I am watching Star Wars and see a spaceship, which is two oblong capsules flying connected in tandem. The Gypsy Shaman says to me telepathically: "Dios es una idea: son dos". I understand that to mean "God is an idea: there are two elements". So I go through life basing my life on the number two.

Chapter 5

Once one has tasted Spain, that person longs to return. I land in Madrid, Northern Spain, The Capitol. The Spaniards know my destination is Granada, Southern Spain, The Gypsy Neighborhood called The Sacromonte, the caves, and immediately recognize I am under the spell of a Gypsy Shaman, and what is more that I am The AE35 Antenna for The Gypsy Space Program. Flamenco being flamenco, the Spaniards do not undo the spell, but reprogram the instructions for me, the AE35 Antenna, so that when I arrive back in the United States, my flamenco will now state their idea of a space program. It was of course, flamenco being flamenco, an attempt to out-do the Gypsy space program.

Chapter 6

I am back in the United States and I am at the house of my mother, it is night time again, she is asleep, and I look out the window east, across the Atlantic, to Spain, and this time I do not see the living room of the gypsy shaman, but the streets of Madrid at night, and all the people, and the word Jupiter comes to mind and I am about to say of course, Jupiter, and The Spanish interrupt and say "Yes, you are right it is the largest planet in the solar system, you are right to consider it, all else will flow from it."

I know ratios, in mathematics are the most interesting subject, like pi, the ratio of the circumference of a circle to its diameter, and the golden ratio, so I consider the ratio of the orbit of Saturn (the second largest planet in the solar system) to the orbit of Jupiter at their closest approaches to The Sun, and find it is nine-fifths (nine compared to five) which divided out is one point eight (1.8).

I then proceed to the next logical step: not ratios, but proportions. A ratio is this compared to that, but a proportion is this is to that as this is to that. So the question is: Saturn is to Jupiter as what is to what? Of course the answer is as Gold is to Silver. Gold is divine; silver is next down on the list. Of course one does not compare a dozen oranges to a half dozen apples, but a dozen of one to a dozen of the other, if one wants to extract any kind of meaning. But atoms of gold and silver are not measured in dozens, but in moles. So I compared a mole of gold to a mole of silver, and I said no way, it is nine-fifths, and Saturn is indeed to Jupiter as Gold is to Silver.

I said to myself: How far does this go? The Shaman's son once told me he was in love with the moon. So I compared the radius of the sun, the distance from its center to its surface to the lunar orbital radius, the distance from the center of the earth to the center of the moon. It was Nine compared to Five again!

Chapter 7

I had found 9/5 was at the crux of the Universe, but for every yin there had to be a yang. Nine fifths was one and eight-tenths of the way around a circle. The one took you back to the beginning which left you with 8 tenths. Now go to eight tenths in the other direction, it is 72 degrees of the 360 degrees in a circle. That is the separation between petals on a five-petaled flower, a most popular arrangement. Indeed life is known to have five-fold symmetry, the physical, like snowflakes, six-fold. Do the algorithm of five-fold symmetry in reverse for six-fold symmetry, and you get the yang to the yin of nine-fifths is five-thirds.

Nine-fifths was in the elements gold to silver, Saturn to Jupiter, Sun to moon. Where was five-thirds? Salt of course. "The Salt Of The Earth" is that which is good, just read Shakespeare's "King Lear". Sodium is the metal component to table salt, Potassium is, aside from being an important fertilizer, the substitute for Sodium, as a metal component to make salt substitute. The molar mass of potassium to sodium is five to three, the yang to the yin of nine-fifths, which is gold to silver. But multiply yin with yang, that is nine-fifths with five-thirds, and you get 3, and the earth is the third planet from the sun.

I thought the crux of the universe must be the difference between nine-fifths and five-thirds. I subtracted the two and got two-fifteenths! Two compared to fifteen! I had bought the Shaman his fifteenth rubber hose, and after he made me into the AE35 Antenna one of his first transmissions to me was: "God Is An Idea: There Are Two Elements".

It is so obvious, the most abundant gas in the Earth Atmosphere is Nitrogen, chemical symbol 15!

Chapter Twelve: The Sequence

We considered the ratio nine to five, then the proportion and found it in Saturn Orbit to Jupiter orbit, Solar Radius to Lunar Orbit, Gold to Silver and if flower petal arrangements. It is left then to consider the whole number multiples of nine-fifths (1.8) or the sequence:

1.8, 3.6, 5.4, 7.2,...

in other words, and we look to see if it is in the solar system and find it is in the following ways:

1.8

Saturn Orbit/Jupiter Orbit
Solar Radius/Lunar Orbit
Gold/Silver

3.6

(10)Mercury Radius/Earth Radius
(10)Mercury Orbit/Earth Orbit

(earth radius)/(moon radius)=
4(degrees in a circle)(moon distance)/(sun distance)
= 3.7 ~ 3.6

There are about as many days in a year as degrees in a circle.

(Volume of Saturn/Volume Of Jupiter)(Volume Of Mars) = 0.37 cubic earth radii
~ 3.6

The latter can be converted to 3.6 by multiplying it by (Earth Mass/Mars Mass) because Earth is about ten times as massive as Mars.

5.4

Jupiter Orbit/Earth Orbit
Saturn Mass/Neptune Mass

7.2

10(Venus Orbit/Earth Orbit)

Chapter 13: The Neptune Equation

If we consider as well the sequence where we begin with five and add nine to each successive term: 5, 14, 23, 32... Then, the structure of the solar system and dynamic elements of the Universe and Nature in general are tied up in the two sequences:

5, 14, 23, 32,...

and

1.8, 3.6, 5.4, 7.2,...

How do we find the connection between the two to localize the pivotal point of the solar system? We take their difference, subtracting respective terms in the second sequence from those in the first sequence to obtain the new sequence:

3.2, 10.4, 17.6, 24.8,...

Which is an arithmetic sequence with common difference of 7.2 meaning it is written

$7.2n - 4 = a\_n$

The a_n is the nth term of the sequence, n is the number of the term in the sequence.

This we notice can be written:

[(Venus-orbit)/(Earth-orbit)][(Earth-mass)/(Mars-mass)]n − (Mars orbital #) = a_n

We have an equation for a sequence that shows the Earth straddled between Venus and Mars. Venus is a failed Earth. Mars promises to be New Earth.

The Mars orbital number is 4. If we want to know what planet in the solar system holds the key to the success of Earth, or to the success of humans, we let n =3 since the Earth is the third planet out from the Sun, in the equation and the result is a_n = 17.6. This means the planet that holds the key is Neptune. It has a mass of 17.23 earth masses, a number very close to our 17.6.

Not only is Neptune the indicated planet, we find it has nearly the same surface gravity as earth and nearly the same inclination to its orbit as earth. Though it is much more massive than earth, it is much larger and therefore less dense. That was why it comes out to have the same surface gravity.

Chapter 14: The Uranus Equation

I asked what needs to be done to solve My Neptune Equation, by going deep with the guitar in Solea Por Buleras. I found the answer was that I didn't have enough information to solve it.

Then I realized I could create the complement of the Neptune equation by looking at the Yang of 5/3, since the Neptune equation came from the Yin of 9/5.

We use the same method as for the Neptune equation:

Start with 8 and add 5 to each additional term (we throw a twist by not starting with 5)

5/3 => 8, 13, 18, 23,...

List the numbers that are whole number multiples of 5/3:

5/3n = 1.7, 3.3, 5, 6.7,...

Subtract respective terms in the second sequence from those in the first:

6.3, 9.7, 13, 16.3,...

This is an arithmetic sequence with common difference 3.3. It can be written:

$(a\_n) = 3 + 3.3n$

This can be wrtten:

Earth Orbital # + (Jupiter Mass/Saturn Mass)n = a_n

Letting n = 3 we find $a\_n = 13$

The closest to this is the mass of Uranus, which is 14.54 earth masses. If Neptune is the Yin planet, then Uranus is the Yang planet. This is interesting because I had found that Uranus and Neptune were different manifestations of the same thing. I had written:

I calculate that though Neptune is more massive than Uranus, its volume is less such that their products are close to equivalent. In math:

N_v = volume of Neptune
N_m = mass of Neptune
U_v = volume of Uranus
U_m = mass of Uranus

$(N\_v)(N\_m) = (U\_v)(U\_m)$

Chapter 15: The Earth Equation

We then sought the Yang of six-fold symmetry because it is typical to physical nature, like snowflakes. We said it was 5/3 since it represents the 120 degree measure of angles in a regular hexagon and we built our universe from there, resulting in the Uranus Integral, which was quite fruitful. Let us, however, think of Yang not as 5/3, but look at the angles between radii of a regular hexagon. We have:

360 – 60 = 300

300 + 360 = 660

660/360 = 11/6

We say Yin is 9/5 and Yang is 11/6 and stick with The Gypsy Shaman's 15 (See An Extraterrestrial Analysis, chapter titled "Gypsy Shamanism And The Universe") and build our Cosmology from there.

We already built The Neptune Equation from 9/5 and used it with 5/3 to derive the planet Europia, but let us apply 11/6 in place of 5/3:

11/6 => 11/6, 11/3, 11/2, 22/3,... = 1.833, 3.667, 5.5, 7.333,...

11/6 => 6, 6+11 = 17, 17+11=28, 28+11=39, ... = 6, 17, 28, 39,...
Subtract the second sequence from the first:
4.167, 13.333, 22.5, 31.667,...
Now we find the common difference between terms in the latter: 9.166, 9.167, 9.167,...

$(a\_n) = a + (n-1)d = 4.167+(n-1)9.167 = 4.167 + 9.167n - 9.167 = 9.167n-5$

Try n=3: 9.167(3) – 5 = 27.501 – 5 = 22.501 (works)
Our equation is:

$(a\_n) = 9.167n -5$

We notice this can be written:

[(Saturn Orbit)/(Earth Orbit)]n – (Jupiter Orbital #) = (a_n)

The Neptune Equation for n=3 gave Neptune masses, the Uranus equation for n=3 gave Uranus masses. This equation for n=3 gives close to the tilt of the Earth (23.5 degrees) in a form that is exactly half of the 45 degrees in a square with its diagonal drawn in. In the spirit of our first cosmology built upon 9/5, 5/3, and 15, we will call this equation The Earth Equation.

Chapter: 16: The Unification Of Pi and Phi by Nine-Fifths

All that is left to do is to consider pi, the circumference of a circle to its diameter, and phi, the golden ratio, since they are two most important, if not most beautiful ratios in mathematics.

I have found nine-fifths occurs throughout nature in the rotation of petals around a a flower for a most popular arrangement, in the orbits of jupiter to saturn in their closest approaches to the sun, in the ratio of the molar masses of gold to silver, and in the ratio of the solar radius to the lunar orbit. I now further go on to say that this nine-fifths unifies the two most important ratios in mathematics pi and the golden ratio (phi), in that

pi + phi = 3.141 + 1.618 = 4.759

Because the numbers after the decimal in the sum (the important part) are 5 and 9 and 7, the average of 5 and nine.

I should also like to point out that the fourth and fifth numbers after the decimal in pi are 5 and 9 and in phi the second and third numbers after the decimal are one and eight where nine-fifths divided out is one point eight, and, further, the first and second numbers after the decimal in phi add up to make 7, the average of 9 and 5, and subtract to make five, and the second and third digits after the decimal add up to nine. So not only does the solar system unify pi and phi through nine-fifths, pi and phi taken alone express nine-fifths in the best possible ways.

Chapter: 16: The Unification Of Pi and Phi by Nine-Fifths

All that is left to do is to consider pi, the circumference of a circle to its diameter, and phi, the golden ratio, since they are two most important, if not most beautiful ratios in mathematics.

I have found nine-fifths occurs throughout nature in the rotation of petals around a a flower for a most popular arrangement, in the orbits of jupiter to saturn in their closest approaches to the sun, in the ratio of the molar masses of gold to silver, and in the ratio of the solar radius to the lunar orbit. I now further go on to say that this nine-fifths unifies the two most important ratios in mathematics pi and the golden ratio (phi), in that

pi + phi = 3.141 + 1.618 = 4.759

Because the numbers after the decimal in the sum (the important part) are 5 and 9 and 7, the average of 5 and nine.

I should also like to point out that the fourth and fifth numbers after the decimal in pi are 5 and 9 and in phi the second and third numbers after the decimal are one and eight where nine-fifths divided out is one point eight, and, further, the first and second numbers after the decimal in phi add up to make 7, the average of 9 and 5, and subtract to make five, and the second and third digits after the decimal add up to nine. So not only does the solar system unify pi and phi through nine-fifths, pi and phi taken alone express nine-fifths in the best possible ways.

Chapter 18: (pi) and (e)

I have talked about how 9/5, which I have found exists in Nature and the Universe, unifies pi and the golden ratio (phi):

(pi) + (phi) = 3.141 + 1.618 = 4.759

because the first three numbers after the decimal are 7, 5 and 9. Seven is the average of nine and five, and the second number is our 5 in nine-fifths and the third number is the 9 in nine-fifths.

It would seem 9/5 unifies euler's number, e, and pi, as well:

(pi) + (e) = 3.141 + 2.718 = 5.859

The second number after the decimal is the 5 in nine-fifths, and the third number after the decimal is the 9 in nine-fifths. The first number after the decimal is eight. This is significant because the 8 is the 8 in 1.8, which is 9/5 divided out. The one will take you all the way around a circle, what is left is 0.8.

Ian Beardsley
January 17, 2013

Data For The Planets

planet	Orbit (O)	Radius (R)	Mass (M)
mercury	0.387099	0.382	0.0558
venus	0.723332	0.949	0.8150
earth	1.000000	1.000	1.0000
mars	1.523691	0.532	0.1074
jupiter	5.202803	11.27	317.893
saturn	9.53884	9.44	95.147
uranus	19.1819	4.10	14.54
neptune	30.0578	3.88	17.23

O for Earth = 1.495979E13 cm   R for Earth = 6,378 km   M for Earth = 5.976E27 g

Earth-Moon Separation: 3.84E10 cm
Solar Radius: 6.9599E10 cm

Molar Mass of Gold: Au = 196.97
Molar Mass of Silver: Ag = 107.87

Saturn (minimum distance from sun) = 9.014 AU = 1.348E9 km
Jupiter (minimum distance from sun) = 4.951 AU = 7.409E8 km

Jupiter (maximum distance from the sun): 5.455 AU ~ 5.4 Astronomical Units

The Yang Elements

K = potassium
Na = Sodium

K/Na = 39.10/22.99 = 1.71

5/3 =1.67 ~ 1.7

K/Na ~ 5/3

We have the Neptune Equation:

7.2x –4

We have the Uranus Equation

3.3x + 3

And now with our alternate cosmology we have The Earth Equation:

9x-5

With three equations we can write the parameterized equations in 3-dimensional space, parameterized in terms of t, for x, y, and z. We can write from that f(x,y,z) and find the gradient vector, or normal to the equation of a plane in other words, and from that a region in space.

$$x(t) = \frac{36}{5}t - 4$$

$$y(t) = \frac{33}{10}t + 3$$

$$z(t) = 9t - 5$$

$$\frac{5x + 20}{36} = \frac{10y - 30}{33} = \frac{z + 5}{9}$$

$$\frac{5}{36}x - \frac{10}{33}y - \frac{1}{9}z + \frac{10}{11} = 0$$

$$\nabla f = \langle 5/36, -10/33, -1/9 \rangle$$

a=5/36   b=-10/33

$$c = \sqrt{(5/36)^2 + (10/33)^2} = \sqrt{0.0918 + 0.019} = 0.3328$$

d=-1/9

$$\tan\alpha = b/a$$

$$\alpha = -65.358°$$

$$\tan\beta = d/c$$

$$\beta = -18.46°$$

-65.358 degrees/15 degrees/hour =-4.3572 hours

24 00 00 – 4.3572 = 19.6428 hours

RA: 19h 38m 34s
Dec: -18 degrees 27 minutes 36 seconds

Ian Beardsley